D1321892

WAKING UP

LIFE ON A SMALL, 'HUMANE' FARM

Vicky Hamill

2QT Limited (Publishing)

First Edition published 2019
2QT Limited (Publishing)
Settle, North Yorkshire BD24 9BZ United Kingdom
Copyright ©Vicky Hamill 2019
The right of Vicky Hamill to be identified as the author
of this work has been asserted by her in accordance with the
Copyright, Designs and Patents Act 1988

Facebook page: @vckhamill

Cover by Charlotte Mouncey
Cover image supplied by iStockphoto.com

Printed in Great Britain by
Lightning Source UK Limited

A CIP catalogue record for this book is available
from the British Library
ISBN 978-1-913071-42-4

Also available as an eBook

To the animals

Chapter 1

1981, Orkney

Geese mate for life. I know that now.

Onion and Sage waddled everywhere together, eating together, sleeping together and washing, oh, so much washing. They had the cleanest, whitest feathers you could ever come across. Often, when the cows came to the water for a drink, they would find these two geese swimming and splashing about in the trough: 'Sorry, no room for cows in here.'

I would call them from across the field and they would start running, a running waddle, towards me, honking, high pitched and joyful. Then, as they came closer, it would turn into a race. They would extend their wings out, flap, flap, flap as they ran, somehow thinking that this would make them go faster. The honking would get more frantic, 'I'm first. No, me, I'm first!' Pushing, shoving and the honking, now a noise more like raucous laughter. Once they reached me, they would get their reward: cabbage leaves or cooked tatties, or a cuddle, each was as good as the other to these guys.

In the spring, Sage made a nest in an old tea chest and started laying eggs, one every couple of days. She had laid about six eggs when she went broody, fluffing out her feathers and covering the huge eggs with her downy blanket to keep them warm. She was a very attentive mum, turning her eggs, talking to them, and she would only leave them a few minutes each day while she ate and had a quick wash. Onion, the good husband, watched over her. We had seen Onion doing his 'job' for a few weeks before so we sat back, excited, waiting for the eggs to hatch.

Three weeks. Any time now. Four weeks... A-ny-time-now... Sage started getting restless, spending more time eating and less sitting... Four and a half weeks... She got up, walked away and left them.

We cracked one open. It was still an egg, a stinking egg, but an egg; no chick. We cracked another and another; no chicks. Onion had been firing blanks. He was not, could not, do

2

his job as a gander.

You cannot carry passengers on a small-holding. This is what I had been told. I didn't like it but I believed it to be true. Everything has to earn its keep.

Be strong.

Onion had his head chopped off ... we ate him.

∞

Sage didn't know where Onion had gone. She wandered around the edge of the field looking for him, calling him. She hardly ate, didn't play in the water trough any more or come running when I called her. She was always by the fence, as if she thought he might be in the next field. First of all she called loudly but after a few days her head went down and she honked quietly, just talking to herself. So sad. She didn't understand why he had left her.

A few weeks passed and then, not a moment too soon, she got a new friend.

We only had three sheep at this point: Lassie, Lola and Lucy. Lassie and Lola were best buddies and Lucy was a bit of an outsider. She was young and in lamb for the first time.

Lucy and Sage struck up a weird and

wonderful friendship. Sheep and goose. A bit tentative at first then, after a few days, they were going everywhere together. They were never apart. Sage and Lucy. Lucy and Sage. Sage started washing again, flapping around in the trough as Lucy tried to take a drink, then Lucy would lie down and Sage would snuggle right into her warm, fleecy coat. She didn't come running any more when I called her but Lucy finally had a buddy and Sage seemed happy again.

The weather was getting warmer. Spring was on its way. Lucy, clever girl, lambed beautifully all by herself. Two beautiful little lambs. She loved them immediately. Sage had kept her distance during the birth and only watched as Mum washed her newborns clean but, once Mum was back on her feet and the babies were trying to work out how to use their legs, Sage thought it was time to say hello to the newcomers. She happily waddled closer.

Lucy's maternal instinct took over and, without hesitation, the new mum chased Sage away. Sage tried again, then again and again; each time Lucy chased her further away from her babies. Finally Sage held back, watching from a distance. Occasionally she would take a few guarded steps towards the new family but a look from Lucy would turn her away again. After an hour or so, she waddled off aimlessly, once again alone and she couldn't understand why.

She wandered out of the field and sat in the middle of the yard. She just sat. Didn't eat, didn't drink and, for me the saddest part, she didn't wash. She and Onion would spend a good part of each day washing and cleaning each individual feather, front and back, then fluff themselves up and strut round, magnificent, showing off. Now Sage just sat there; her once-beautiful white feathers were yellow and full of dirt and she didn't care. I put a bucket of water right next to her and all of the foods she liked. She didn't move. Just sat. I put food and water right into her beak. She let it fall out. I called the vet. He didn't know what to do; there was nothing 'wrong' with her.

She sat... Just sat ... and died.

Did you know you can die of a broken heart? Sage, a goose no different to any other goose, died of a broken heart.

2

Born Vegan

I think I was born vegan, I just didn't know it. I was certainly born an animal lover. I wouldn't be surprised if my first words were, 'I want a dog.' I had animal toys, animal books, animal bedcovers and animal wallpaper. Then came the day that my dad brought home Tiny.

Tiny was a kitten, an adorable ball of cuddly, playful fluff.

I would have been about five or six years old, living with my family in Gloucester. Dad came home from work one evening carrying a gorgeous black kitten who had been born to one of the factory cats which they kept to keep the mice down. Dad handed him over to me and said I could keep him.

I was beside myself with joy. I didn't know

what to do! My own kitten. I would love him forever. I would look after him. I would never leave him. He was Tiny, black with one white paw and a white splodge just above his little wet nose. We played all evening and at bedtime Tiny came up to bed with me and snuggled in close, purring happily under the blankets. That, at the time, was the happiest night's sleep of my life. I dreamed about nothing but Tiny and when I woke he was still cuddled in close.

By the morning my brother, Danny, who was five years older than me, was having a full-blown asthma attack. The doctor was at the house and my mum was in a panic. Tiny, who was the deemed cause of the attack, was scooped up and promptly sent back to the factory. Heartbroken, I cried and cried. I screamed. I begged. My world fell apart. I remember sobbing, 'Why do we have to get rid of Tiny? Why can't we get rid of Danny?'

Of course we didn't get rid of Danny, and Tiny never came back. I begged and begged, first of all for Tiny but then for another pet, and another, and so began a string of pets: a dog, guinea pigs, goldfish, mice. Some that Mum knew about and some that I smuggled into my room.

During this time I started to realise that meat came from animals and I announced to my parents that I didn't want to eat meat any more. Mum and Dad were used to me having these crazy ideas regarding animals; I wouldn't even squash a bee! So, unfazed, Dad calmly informed me that we have to eat meat to live. I think now that he truly believed that. He told me that animals live happy lives on the farm (I think he believed that, too), and that we ate them when they died. I'm not so sure he believed that, but I did and for a long time that is what I believed.

One day, when I was about eleven, I went up to my room. The window had been left wide open and there, wandering around the floor, was a pigeon. I slid over to the window, shut it and went downstairs to get food and water. Sitting on the floor, I quietly offered the food to the wee chap and in no time he was eating happily out of my hand. Of course, I knew I had to let him go so, after I had spent some time with him, I opened the window again and stood back. He hopped up onto the sill, appeared to say his goodbyes and flew off. I waved goodbye.

I was happy; I had spent a few special moments with one of Nature's animals. So imagine my delight when he came back. Not once but every day, back he came. I would leave the window open for him if I could but, if he found it shut, he would tap on the glass for me to let him in. I called him Dilly. Often, Dilly would fly over close to me in the garden, sometimes landing on the lawn next to me. I started leaving my window open at night and he would come in and sleep on top of my wardrobe.

Then one morning we found his little body on the ground below my window. Mum said that maybe he had fallen off the windowsill. Parents say all sorts of things to their children and the child completely believes them. It was only later that I thought, wait a minute ... birds can fly! I can only guess that he was old and had found an easy, comfy life for his last few weeks; hopefully he died of old age in his sleep.

By the time I went to secondary school, my only pet was a beautiful golden retriever called Sultan. I missed him when I was at school and rushed home to him afterwards, spending my weekday evenings with him and my weekends helping out at the local riding stables in exchange for free rides.

∞

I truly believed that my pets, indeed *all* pets and horses and farm animals, were living happy lives and I knew that I wanted to live and work with animals when I left school. I also knew that elsewhere there was animal cruelty going on. I was sickened by seal culling, whaling and vivisection; I was horrified when the boys down the street pulled the legs off daddy long legs, and I wouldn't watch horse racing or cowboy films where the horses fell or were pulled over. My dad, again not wanting to see me upset, told me that the horses were trained to do that. While I wanted to believe him – he was my dad after all – I couldn't see how, trained or not, those horses didn't hurt themselves as they crashed into the ground over and over again.

So I became an activist against these things, just not a very good one. I never even considered that one ordinary schoolgirl could do anything that would make a change. At school, I set up the Caring Club for Animals, the CCFA, and a lot of my classmates joined. We wore handmade badges and made leaflets. We had little get-togethers and talked about animals and animal cruelty. We never did anything productive but I guess our hearts were in the right place and if someone would have led us we would have followed.

I wonder if any of those girls are veggie or vegan now, or did regular, conventional life take over? Did they go ahead and follow the norm as most people do?

3

The Flock of One

When I left school, I got a job at a riding stables and trained as an instructor. I guess this is when I started to get hard. I got used to country life and the practices that were deemed normal for looking after horses and other animals. Some things didn't sit well with me but who was I to question them? After all, they had always been done that way.

I met my then husband in Dorset and we lived there briefly. He was a keen gardener and wanted to live off the land. My dream was to have a goat for milk and some chickens for eggs. So in October 1979, unable to afford property in Dorset, we took the plunge and moved to an island in Orkney, having bought a smallholding of a couple of acres with a few outbuildings. We piled into a van with our two dogs and all our worldly goods and drove from Dorset to Orkney, a long, cramped, uncomfortable ride. It was wonderful. The adventure was beginning.

∞

Shortly after moving into our new home on the small island, some of the locals appeared, keen to meet the new incomers. They were lovely. They brought us tatties, neeps (swedes), homebrew; one even gave us a few chickens and a cockerel (to get us started) and two kittens (to keep the mice down). One of the incomers sold us a goat, Schnooks, already in kid. So within a few days of arriving I had my goat and chickens. Life could surely get no better.

One of the local farmers invited us round to his farm. We had tea and cakes and a wee dram, which was traditional whenever someone came into your house, and then he showed us round his farm. The cows had been brought inside for the winter and were in a huge barn. They were in individual stalls with chains around their necks, but I barely looked at them. As we walked in, my attention was immediately drawn to a bloody, severed pig's head lying on a bench at the door. Its eyes were wide, as if in horror, its mouth open and its swollen pink tongue was bulging out.

The farmer saw me looking at it. Picking it up, amused, he thrust it towards me. Did I want to take it home to make potted heed (potted head)? I must have reeled away from it. Er, no thank you. I didn't think I was quite ready to make potted heed just yet. The farmer laughed and tossed the head back on the bench. It was like something from a horror movie. I wondered how anyone could get hardened to things like that.

It must have been about a week later when he asked us if we wanted to buy some mutton. We had been eating just neeps and tatties with the odd egg, so we said yes. A bit of meat in the freezer would be good. Did we want a half of mutton or a whole mutton? We would have to cut it up ourselves. The freezer was empty so we decided to take the whole mutton. The next day the farmer pulled up in his Land Rover with a very live sheep in the back. He stepped out, a big smile across his rugged face. Did we want him to kill her for us?

I think there was a long, shocked silence. Never before had we faced the fact that in order to eat meat we would need to kill the animal. And there she was, looking at us from the back of the Land Rover, dirty white fleece, black head and big brown eyes. If we did want him to kill her, he could do it then and there. He had brought a sharpened knife, it wouldn't take a minute. All I could think of was the bloody, severed head of the pig I'd seen back at his farm. Suddenly mutton didn't sound so appealing.

We had never planned on having sheep but now we had a very much alive flock of one. Our freezer was still empty but I was happy. The farmer took the ewe away again, ran her with his ram and returned her, in lamb, a couple of weeks later. I called her Lassie. She was only three years old, had lambed twice, but was now considered mutton and had been sentenced to death. Her crime? At each lambing she had only produced one lamb. This was enough reason for the farmer to cull her and replace her with a ewe that would give him twins.

The farmer was a nice guy. He wasn't doing anything considered cruel or uncaring. This was just normal farming practice.

By our first spring in Orkney we had:
two dogs, Sultan and Pepsi
two cats, Bonny and Clyde

two goats in kid, Schnooks, and another goat, Jezz

one ewe in lamb, Lassie

around a dozen chickens

... and a cockerel.

4

My First Kid

Schnooks was the first to kid. She was a Golden Guernsey cross with a beautiful long golden coat and a very clever mind of her own. I noticed that she was starting to develop an udder; over a few weeks it got bigger and, with that, I started to get nervous. My first kidding. I hoped she knew what to do because I had no idea. If I thought something wasn't right, my plan was to run across the field to get my neighbour.

I checked her regularly and one day noticed that she was pawing the ground. She would turn her head to her belly and gently bleat as if talking to it. She looked unsettled, lying down, getting up and turning round. Then she walked over to the wall and stood with her head firmly pressed against it, not moving. Er, was that normal?

That's when I noticed the contractions. My heart started pounding. Should I do something? How long should this take? When do I know something is wrong? When do I start running?

Schnooks did know what she was doing and it didn't take long before a nose and two tiny front feet slid out, followed by a little wet head, shoulders, body. And then, plop, a whole kid flopped onto the floor, still encased in a membrane and covered in yellow goo. Schnooks turned immediately, talking to it, 'maa, maa, maa'. She washed it, head first, and as its head became free of goo, it lifted it up and started talking back with tiny bleats. Schnooks was now apparently on a mission to have the cleanest kid in the world.

Of course at this point, I was sobbing loudly, tears of joy and wonder at what I had just seen. I gently lifted a back leg and had a peek. It was a little boy, a beautiful little boy. When every drop of yellow goo and every trace of membrane was gone, Schnooks was still going. Talking as she licked, she washed behind his ears, under his tail, up his nose, everywhere.

By now he was making his first attempts at standing but he'd only get his front legs under him before the licking knocked him over again. His little bleat got louder and a bit frantic; Okay, Mum, you can stop now. I'm clean.

But Mum was still on her washing mission. Mum knew best.

I left them to it and went in for a coffee.

Tyler. His name was Tyler.

Half an hour later, I went to see how Schnooks and Tyler were getting on. Mum was lying down having a wee rest but she was still close, still licking, although now a lot more casually. Tyler was drying out and starting to get fluffy.

I know that he was my first kid, and I know I'm completely biased, but there has never been – nor will there ever be – a cuter, more beautiful, more perfect kid than Tyler. His mum was golden and his dad was a British Alpine, black with lovely white facial markings, white legs and under the tail. Tyler had the best of both: perfect British Alpine markings but in deep gold and soft cream. He was just gorgeous.

However, cute as he was, he hadn't mastered standing. He could get his front legs under him but the back legs weren't playing ball. I decided to help and lifted him gently to his feet. Mum stood up too. His back legs still weren't holding his weight so I held him to his mum's teat. He latched on immediately and drank furiously. I could feel his little tummy getting full. I laid him back down and left them again. He would soon be strong enough to stand on his own.

After a day of holding him up to drink from his mum, I started bottle feeding him. I milked his mum; we drank her milk while I fed him bought powdered milk. At three days old he seemed to be fit, strong, healthy, even happy, and his little tail wagged frantically as I fed him. But he couldn't stand up; his back legs just didn't work.

My husband and I had huge arguments. I was determined to care for Tyler until he could stand. I could clean him (he was constantly lying in his own pee because he couldn't move), feed him, move him... We could make him a little trolley for his back end...

You cannot carry passengers on a smallholding. He was male and, even if he had been healthy, he was of no use. We were wasting money on him. I took him to the vet twice (which almost caused a divorce then and there) but the vet agreed; Tyler was paralysed from half-way down his back and there was nothing that could be done.

We had to toughen up if we were going to live this life. My husband would kill him. Hit him over the head. I couldn't bear it. I took Tyler to the vet a third time and held him close as he was quietly put to sleep.

5

Kids , Lambs and Piglets

Our other goat, Jezz, was pure white. A Saanen. The lady who owned her wanted rid of her because she was getting another goat. Jezz had a horn, just one because the other had been broken off somewhere, somehow, and the lady didn't want her attacking her new goat with her one vicious horn. So I took her. Jezz turned out to be a beautiful gentle soul who showed no signs of aggression to anything and the horn was never a problem.

She was already in kid and that first spring she produced two delightful white offspring. My heart sank a little when I saw that they were both male. I knew what their fate would be but, for the time being, they were going to live.

They spent their first day with their mum, suckling so that they could get the first milk, and Jezz loved them. She washed them, nuzzled them as they fed, talked to them with soft, low vibrating bleats, and at night they all cuddled up together, a happy little family. On day two her babies were taken from her. I had to take them – we wanted the milk. That's why we had goats, meat and milk. Everything had to earn its keep.

I lifted them, one under each arm. Jezz, immediately distressed, stuck super-close, her nose next to their dangling back legs, and she called hysterically to them. I managed to get out of the door and shut her in. If a goat can scream, she was now screaming and bashing at the door.

The kids were so helpless, just hanging there as I carried them, calling back to Mum for help. I put them in a separate barn where they stood close to each other, bleating for their mum. I put Jezz, along with Schnooks, out into a field. I expected them to trot off and start eating as they normally would, but Schnooks was still missing Tyler and she and Jezz ran up and down, up and down, calling for their kids.

It hadn't occurred to me that they would be broken-hearted. They had just lost their babies. They would be sad, I thought, yes, but they'd get over it. It must have been worse for Jezz because she could hear her babies in the barn calling back. Or maybe it was worse for Schnooks because she could hear Jezz's babies calling back to their mum; this meant they were alive, they were close, but her baby was just ... gone.

I went back into the house where I couldn't hear them calling. That had been a bit more traumatic than I thought it would be, but these things had to be done and I'd done it. I was a real smallholder.

Jezz's babies were hungry and I started bottle feeding them every few hours. After a couple of days, while Jezz was still dripping with milk for them every morning and constantly calling, they started to look at me as their mum.

∞

The local farmers were all very friendly and helpful. We would regularly ask for their advice and they were happy to give it; happy to tell us how things are done on the farm, how things had always been done. So we became very friendly with a few farming families.

One of the farmers asked me if I would take a wee lamb. Her own mother had rejected her; he had tried to foster her unsuccessfully onto another ewe, and his wife had neither the time nor the inclination to bottle feed her. Basically, if I didn't take her, her future would be grim.

Of course I took her. I named her Lola and added her to my little gang of babies: two kids and now a lamb. I cannot tell you what a delight bottle feeding is. There is pure excitement at every feed as they punch the bottles with their little noses, froth foaming out of their mouths, tails wagging madly. Given the chance, the babies would follow me everywhere, sit with me, climb all over me; they were the embodiment of joy. The kids especially loved climbing; they would run and play, frolic and climb, climb on everything. They could get up onto the windowsills of the cottage and take a flying leap off with a little flick through the air like a BMX rider. Then they would come running back to me, pleased as punch. 'Did you see me? Did you see what I did?'

∞

One fine day Lassie, the 'old' ewe, went off into the corner of the field by herself and settled down to have her lamb. I kept an eye on her from a distance but she didn't need any help. She produced one big healthy lamb that we called Lucy. We now had a flock of two, plus the bottle-fed lamb who, at the moment, thought she was a goat.

The wild forebears of sheep naturally had short tails. Over time, farmers selectively bred sheep to have more fat on them and a consequence of this is that their tails have grown longer. Now, having bred them to have long tails, farmers cut them off at a few days old. I didn't realise this. Lola had already had hers cut off when we got her so it was a surprise to me when Lucy was born with a long tail. I honestly thought sheep were born with short tails. Anyway, there was no way that I was going to cut off Lucy's tail so she lived that summer and, indeed, the rest of her life with a long tail. It didn't get in the way of anything, it didn't develop any nasty diseases; in fact I didn't know then, and I still don't know now, why sheep have their tails cut off.

Spring is a busy time. One of our farming neighbours asked if we wanted any piglets; he was ordering some from mainland Orkney to grow through the summer and to put in the freezer around October. We still had nothing in our freezer. It had been a long winter living on neeps, tatties, eggs and the occasional chicken, so we said yes.

The piglets arrived on the island boat in a little crate; there were four of them, two for the farmer and two for us. A harbourman swiftly grabbed them by their back legs and lifted them out of the crate. No time for niceties. They were about six weeks old, just scared babies. He handed us the two squealing, wriggling piglets, a boy and a girl. We each took one in our arms and put them in the back of our car. It was only a five-minute journey back to the croft but they stood shaking in the car, grunting, squealing and pooping.

Once home, I put them into their little hut on a bed of straw and sat quietly with them. Still shaking, they stood in the corner for a while but curiosity soon got the better of them and they tentatively came over to give me a sniff. Initially when I moved a hand towards them they squealed and backed away, but it wasn't long before they were standing having their backs scratched or rolling over for a tummy rub. They showed trust in me very quickly and were keen to be friends.

I loved their little flat noses that rooted through the straw and got into everything, and their little waggly tails. They were lovely little characters, everything that the farmyard piglets in the children's books of my youth had been. They had, however, already had their teeth clipped, their tails cut short and the male had been castrated, all when they were just a few days old and all without anaesthetic. But that was how these things were done and now, at least, that was behind them. Now nothing bad was going to happen to them ... for a short while anyway.

By the end of this spring we had added:

- Jezz's two male kids
- one bottle-fed lamb, Lola
- Lassie's lamb, Lucy
- a pair of geese, Sage and Onion
- a young sheepdog, Ben

... and two piglets, Danish and Petunia.

6

Summer

Our chickens were completely free range. I didn't even shut them in at night; there were no wild predators on the island so the chooks went in and out as they pleased and could wander as far as they wanted. Under normal circumstances they didn't wander far at all but, as spring moved on, I noticed that one would disappear. I didn't worry. After a couple of weeks she would be back with a little clutch of chicks following behind.

A hen makes the best mum. While sitting, she turns her eggs numerous times a day and, as she may have ten or more eggs, she is basically attending to them all the time. She chirps to the chicks in the eggs and, a few days before they hatch, they start chirping back.

They soon know each others' chirps so they are bonding even before they have hatched.

Once hatched, the hen's life revolves completely around the chicks but she looks so happy and content; this is what she was born to do. She breaks up food for them and encourages them to eat it with little chirping noises, teaches them to scratch and takes them to water, talking to them all the time. Different chirps for different things. If she sits and sunbathes for a while, her chicks will climb all over her and she doesn't mind. And when they are sleepy, they crawl underneath her warm, downy feathers. She fluffs up to accommodate all the chicks and fuzzy yellow heads poke out from the most unexpected places. She is the bravest of mothers; at any sign of danger she will either call the chicks under her wings – and they obey instantly – or she will chase the danger away, hackles up to make herself look big, wings extended wide. She will take on anything to save her chicks.

Through the spring and summer a few hens would sneak off. I took the eggs from the nests that I found; we didn't want too many hens going broody. This occasionally had unpleasant consequences as I would crack an egg open into the pan and a dead, half-formed chick would drop out. It's easy to forget that that's what an egg actually is. I would retch and suddenly my appetite would be gone, especially for eggs.

Other hens, the ones with the better hiding places, would disappear and then come back with their little families, looking very pleased with themselves, and so our flock of chickens grew. We kept the females but the males, once big enough, were killed and put in the freezer. My husband did it. I didn't watch and I never asked about it, although I knew what he did: neck dislocation. I stayed in the house. Out of sight, out of mind. Once they were dead, it was my job to pluck and disembowel them.

∞

The summer days are long in Orkney and, although it still called for a few layers of clothing, the weather was fair. The animals seemed to be happy and healthy and my husband and I spent happy hours, happy days, outside – him with his crops and me with my animals. We were living off the land. We were smallholders.

The bottle-feds, kids and lamb, were soon big enough to be weaned and I put them into the field. However, they were small enough to get through the fence and they would still come into the house at every opportunity ... as would the chickens.

Directly outside the front door was a really sheltered spot, quite a sun trap. Sultan, the Golden Retriever who was now getting on in years, loved nothing better than lying there. Chickens love sunbathing too and one of the young roosters discovered that the best place was on Sultan's back; warmth from the sun from above and warmth from Sultan below. Not wanting to disturb him, Sultan would not move until the wee guy got off.

In the evening, when the other chickens returned to their barn, this wee rooster would come into the porch and snuggle up with the three dogs for the night.

∞

The pen that the pigs were in was somewhat small, and Danish and Petunia were both getting quite big. They were intelligent animals. In a natural environment they would have been running around, exploring, digging up roots, rolling, sunbathing and interacting with their family group. Here in their hut and pen they must have been painfully bored: walk into the hut, walk out, lie down, stand up.

To amuse themselves, they would rattle stones in the dyke that surrounded them. My husband, hearing the noise, would rush out thinking that they were destroying his dyke. I swear the pigs would laugh as he loomed over the wall, cursing them; their lips would curl up into a big smile and they would grunt a laugh. My husband would check that no damage had been done and leave them, muttering to himself. They would always wait and, just as he was back inside the house, they would do it again. Out he would go, calling them all new names. It was their game.

The only way I could stop them from doing it over and over again (although I would let them do it a few times as, unlike my husband, I thought it was hilarious) was to go and sit in with them. Pigs are naturally very clean animals – they will never pee or poo on their bed – so I could confidently go in and sit on the clean straw and scratch their faces, backs and bellies while they sniffed around my neck and through my hair with their big, soft noses, grunting happily and telling me stories. If I stopped the scratching they would 'paw' me, like a dog hankering for attention. In those moments they were happy ... and so was I. This was what it was all about.

∞

We had rented another two acres next to our smallholding. We used two acres to graze the animals and let one acre grow to make hay for winter feed. The final acre housed the cottage and outbuildings, a small vegetable garden and an area where we grew tatties. Our little vegetable garden grew all the veg that we needed for the whole year so the other three acres went to feed the animals. On top of that, we bought in grain and goat mix for animal feed, dried milk powder for bottle feeding, together with vitamin and mineral supplements and straw for bedding. The four acres was too small to support even the few animals that we had.

7

Kill Time

Days were getting shorter. I had spent a wonderful summer bottle feeding and playing with my babies. The kids were getting more boisterous, playing butting and chasing games, but even with their little horns they would never hurt each other or me. Lola Lamb stayed my baby. She would go out to graze with the other two sheep, Lassie and Lucy, and although she became good friends with Lassie, she would still come running over to me whenever she saw me and nudge me for a head scratch.

The hay was in. The tatties were up. It was time. I didn't want it to come but it came anyway.

October. Kill time.

Of the two pigs, Danish was bigger than

Petunia so he was going to 'go' first, together with Jezz's two kids. We would let Petunia grow on a bit more. Danish would be the first big animal that we had killed. I say 'we'; I was, again, going to hide in the house.

In my head, as soon as the animal was dead it would no longer be the animal I knew; it would be meat. I told myself that the animals had lived happy lives and we wanted to eat meat, so it would be 'happy meat'. I told myself that they weren't going through the trauma of being sent to a slaughterhouse. To make it easier, the ex-slaughterman for the island had offered to loan us his bolt gun. So this would definitely be the best way to do it. The humane way.

The day came. My entire body was horribly numb. I had got Danish as a piglet. I had been there when the two kids were born. I had fed them, played with them, scratched their tummies and curled up and laid with them. They trusted me. They were six months old.

I had already starved them for twenty-four hours so that their insides were clean and now I led them, one by one, round to the barn where my husband was waiting. I handed them over to him, deserted them without a word or a backward look, and ran back inside the house.

I stood and looked out of the window to the field where they had been playing just that morning and I prayed that everything would go alright.

I heard the loud bang of the bolt gun followed by silence. A little time passed; another bang. I knew he was 'doing' the kids first. That must be them dead now. I waited, heart thumping, for the third bang.

Bang! There it was, followed by frantic, high-pitched squealing, screeching, screaming. My heart exploded inside me. I started for the door and hesitated. My world was filled with the distraught squealing; my whole being was telling me to go to Danish, to save him, but I was so scared of what I might find.

Bang! I jumped... Silence. I waited. Still silence. It was done. I collapsed to the floor, sobbing.

The next time I saw them they were carcasses. I laid them on the kitchen table and cut their flesh into chunks, bagged it and put it in the freezer. Outside, my husband washed down the floor of the barn and swept the blood away.

I knew the process was horrible but it never entered my conscience that it might be wrong. In fact, I believed that we were doing the right thing. It was either this or buy our meat pre-packed from the shop and not think about the factory farm that it had come from or the industrial slaughterhouse where it had died.

This was the best way. We were doing the right thing.

8

Babies All Round

The winter was cosy. There was socialising, roaring fires and homebrew, and we had way more than enough to eat. The animals had shelter, inside and out, beds of straw and plenty of hay and feed. Jezz and Schnooks, now in kid again to a neighbour's billy, were still milking well. Although the chook's production had slowed down a little, we were still getting a good few eggs.

Lassie, Lucy and Lola Lamb had gone 'on their holidays' to visit a neighbour's ram. Lola was very nervous as she went away in the Land Rover and called for me all the way as it disappeared down the road. She was still getting used to the idea that she was a sheep. When they returned a couple of weeks later, she stuck by my side again, following me everywhere. It was like, 'I'm not leaving Mum again. Bad things happen when I do.' She was so young. I made a promise to myself never to put a sheep to the ram until their second year.

I bought another goat, Alex. She was a pure bred British Alpine, in kid to a pure bred. I thought I might become a BA breeder, that way I could sell the male kids instead of having to kill them and we would make some extra cash. We also bought a pair of geese, Sage and Onion. We might also make some money from goose eggs and goslings.

Our second spring was on its way.

I felt sure that our animals lived a good life and, although it was hard, we decided that if we didn't raise the animal and kill it ourselves then we wouldn't eat meat. We would only eat our 'happy meat'. For our first Christmas we had bought a frozen turkey but now, in preparation for the next Christmas, we got a day old gosling chick, Busby.

Geese, like hens, are amazing mothers. They bond with the gosling while it is still in the egg; once hatched, they teach their babies everything, so chicks born in a hatchery with no mums are just lost little souls. When Busby arrived by himself in a small closed box, he had no idea what was going on. He just wanted comforting. When I lifted him, he didn't try to jump out of my hands; he seemed to like the warmth and the contact. He wanted cuddles.

I put him in a big open box with bedding, food and water and a little heat lamp but, as I tried to leave him, he started jumping up the sides of the box, cheeping loudly, calling me back. He knocked his food and water over as he waddled around the edges, little wings wide helping to keep his balance, trying to find a way out. I left him for a while to see if he would settle but he just got louder, so I took him out of the box and put him on the floor. He climbed up and sat on my feet. From then on he followed me everywhere like a very small dog. It was comic. Wherever I went, there was Busby, waddling a few paces behind trying to keep up.

His walk was somehow connected to his cheep, or so it seemed, like one of those little remote-control dogs that walks along and yaps. I always knew when he was behind me, 'Cheep, cheep, cheep'. Don't step back. Whenever I stood still, he would climb back onto my feet; when I sat down he would do little jumps, asking to come up onto my knee or, even better, onto my shoulder, where he would sit happily nibbling my ear until he fell asleep, purring like a kitten. We spent many a sunny evening sunbathing behind a dyke, me lying on the grass, Busby snuggled up on top of me.

We were given another young goose, Parsley. She was older than Busby, already feathered up. Although she stayed close to our main pair, Sage and Onion, and they didn't mind, she wasn't in their gang. They were an item and they weren't interested in anyone else. Parsley just tagged along.

As Busby got older he gained a little confidence and would go off into the field to graze with Sage, Onion and Parsley, but he regularly came waddling back to me for a cuddle and a chat.

∞

Lassie was first to lamb that spring. Good old reliable Lassie. She went off into a corner of the field by herself and produced her one lamb. No fuss, no bother. I took a towel to the lamb and, as Lassie licked her baby, I helped dry it off and made sure any mucus was clear of its mouth. It shook its head and did a little bleat. Lassie bleated back happily. I had a little peek ... it was a girl. I smiled. We would keep her in the flock and her name would be Morag.

Lucy produced two big beautiful lambs, a boy, Mac, and a girl, Mona, and all three were fine and well.

This only left my young bottle-fed, Lola. I was in the kitchen when I heard Lola's familiar bleat, but this time it seemed desperate. I went out and she was at the fence, calling me. I could see that she had discharge coming from under her tail. She was about to lamb but, instead of separating herself and going somewhere quiet, she had come to find me.

I tried to get behind her to see what was going on but she just kept turning around with me, bleating, panicking, asking for help. I knew there was something wrong but I couldn't get round her; she kept her head at me all the time, practically climbing into my arms. It was time to do the run.

I jumped over the fence and sprinted across the field to my neighbour. Lola ran up and down the fence calling after me. Two minutes later we were back. I wanted my neighbour to hold her still so that I could get round to her tail end. I had been on a lambing course a few weeks earlier so I knew what to do, but practising on a box with a hole cut in it was not the same as the real thing.

I inserted a shaking hand inside. There ... I could feel something... What was it?... A nose. A little wet nose. There's its head and ears. Now there's a leg on one side – and on the other side? Nothing. I had to push in a little bit further, find the leg ... there it was. It was folded at the knee.

I was so nervous. I was giving a running commentary about everything I was doing, everything I could feel, supposedly to my neighbour but really to myself to ease my panic. I tucked a finger round the back of the lamb's knee and straightened the leg out. Lola gave one big push and out the lamb slid. A beautiful, wet, slimy, black lamb. A miracle.

In my running commentary I had said out loud, 'Oh please, let it be alive. Please let it be alive.' For a couple of seconds, the longest seconds of my life, it just lay there. I wiped the membrane off its nose and face and prayed. It lifted its head, shook and gave a little bleat. Lola bleated back. Once again I burst into tears. This time tears of utter joy.

9

Soldering Irons and Elastrators

Lucy had taken to mothering straight away, chasing off anything that came near her lambs (including poor Sage, who had recently lost Onion). Lola, though, was still a mummy's girl and needed a bit of help. She started licking her baby girl clean; she nuzzled her and talked to her, and then she'd lift her head and talk to me, asking me if she was doing okay.

The baby, who we called Mandy, was soon struggling to her feet. It seems to be common with lots of animals that the mother is so keen on licking that, when the baby finally succeeds in getting to their feet, Mum knocks them back down again. Mandy, after numerous attempts, finally managed to stay up and withstand the cleaning regime. She took unsteady steps towards her mum's udder but Licking Lola followed her around, and the teats just kept moving away. I watched for a while, not wanting to interfere too soon, all the while applauding Mandy's determination. Round and round she staggered, falling, getting back up and trying again. Lola would not stand still. Eventually Mandy fell down; her newborn legs, having done so much, gave up. She stayed down, resting her little nose on her knees, and complained quietly to herself.

It was time for me to help out. I congratulated Mum on her excellent cleaning skills and told her that it would be a good idea to let her baby feed now. With one arm around Lola to keep her still, I lifted Mandy and held her nose close to the teats. Lola could easily have struggled and moved away but she didn't; I think she knew I was helping.

Mandy seemed to know what was happening too; her tail wagging excitedly, she nudged the udder a few times with her nose, latched on to a teat and sucked. Her whole body wriggled with glee, her lovely long tail went into overdrive and she was suddenly full of energy again. Lola realised that she could turn her head and lick her baby's rear end without moving away the milk feed and, although she still knocked Mandy over a few times, they both finally seemed to know what they were doing. Everyone was happy. My job was done.

Now it was the goats' turn. Schnooks and Jezz kidded with no problems and this time, Schnooks' baby girl was fit and healthy. Schnooks was very pleased with herself. Jezz produced two delightful white kids, a boy and a girl. Both goats were proud, loving mums ... for the whole day that they got to be mums. I was dreading day two. I hoped it would be better this time.

We wanted the milk. Day two ... I took their babies away.

It was no better.

Our plot of land was small. There was nowhere I could move the kids to where the mums and babies couldn't hear each other's calls. This time I left the mothers in the barn so that I couldn't hear them, at least in the house. The only way to handle their frantic calling was to block it out, to turn up the radio.

∞

My new goat, Alex, was a lady. Jezz was the matriarch, like the grandmother who keeps everyone in their place with just a look, and Schnooks was a tomboy who needed keeping in her place. But Alex, she was a real lady. She was taller than the others, fine and elegant. She loved a fuss but always made it look like she was doing you a favour in allowing you to stroke her, and she talked a lot about very important things.

Alex kidded with no problems, two stunning babies, a boy and a girl. This was perfect. I could keep them both: the girl, Donna, as a milker and the boy, Jamie, as a stud male. It was the start of my British Alpine goat herd. Hard to imagine because, right then, they just looked like two adorable, long-legged puppies nudging their noses up into Mum's udder, bleating the equivalent of 'nom, nom, nom'. Mum gently bleated back, 'Now, now children. We don't talk with our mouths full.' They were a beautiful, contented family.

Day two... We wanted the milk. I took her babies away.

∞

Our farm was full of babies: four lambs born

on the premises, a new bottle-fed lamb, Millie, who we had been given, and five goat kids all playing, jumping, bucking, climbing and running round in little gangs.

Within the first week of their lives, the lambs needed their tails docked; the local farmers and all the information we had told us that we needed to do this. Little Mac, Lucy's lamb, needed to be castrated and all the kids needed disbudding to stop them growing horns. Goat keepers with much more experience than us agreed – it had to be done.

The castrations could be done either:

- with a knife: cut through the scrotum and then pull the testes out
- by crushing the cords: cutting off the blood supply to the testes causing them to atrophy
- by a rubber ring being put around the scrotum above the testes that is so tight it cuts off the blood supply and causes the whole thing to drop off after a couple of weeks.

The majority of farmers on the island used the rubber-band method for castrations. Tail docking could be done with a knife, by crushing or with the rubber band. We decided on the rubber-band method for both.

I didn't need to drive the sheep into an enclosed area to catch the lambs; I could do it outside because their mums all trusted me and so did the lambs. I slid a tiny but super-strong band onto the four prongs of the Elastrator and set it down ready.

Lucy didn't move away as I approached and I picked up her little boy easily. I knelt down and sat him on his bum, laying him back onto my knees so that he was exposed to the world. Holding him there with one hand, I picked up the Elastrator and squeezed the handles together. The rubber band stretched open. Mum was happily grazing. I slipped the open band over his scrotum and checked the testes were both in place. I took a deep breath. I knew I had to be controlled, but quick. 'Don't drag it out.'

I released the handles. The band tightened but the prongs were still trapped inside. I needed to wriggle them out of the tight band. The lamb cried out loudly and started to struggle, his back legs kicking frantically. Mum came over and trotted worried circles around me, bleating back to him. I pulled at the Elastrator. The prongs dragged out and the band snapped to its tightest. I checked again that I'd done it correctly and set him down on his feet. Little Mac cried even louder and tried to run to Lucy but I couldn't let him go to her. I still had to do his tail. 'Just get it done.'

Another band over the prongs, slide it up his tail, release the handles, wriggle the prongs out and let him go. He bleated, loud and pathetic, flicking his tail. With back legs held wide, and suddenly uncoordinated, he staggered over to his mum. I can only guess at the pain he was in. He immediately went for a drink, looking for comfort, hoping it would make it better. While Lucy consoled him, I picked up Mona and did her tail ... then Morag's ... and then Mandy's. I went round to my little bottle-fed, Millie. She came running up to me, wagging her long tail. I slipped the Elastrator over it...

There was no playing or running in the field that afternoon. The babies lay down and stood up over and over again. They would take short drinks from their mums but then suddenly stretch their necks out and curl up their top lips. They would bleat little cries for help, kicking out with a back leg. They snuggled in as close as they could to their mums and tried to sleep it off, but it wasn't long before they were standing up and kicking out again.

I had intended to castrate Jezz's little boy, but I couldn't bring myself to do another castration so I left him. He was destined for the freezer anyway. I decided not to disbud him for the same reason. At the time, he seemed to be the lucky one.

The rest of the kids took a trip to the vet for disbudding. They were just a week old, happy, lively and mischievous. They had no idea what was coming. One at a time the vet anaesthetised them. He checked they were 'sleeping' by touching an eyeball ... there was absolutely no reaction. It was like they were dead.

He had heated up a tool like a big soldering iron and he held it onto one of the little horn buds on the top of their heads, pressing it in and twisting. The smell of burnt hair and flesh was intense but, worse, every one of the kids even in their unconscious state cried out a loud soul-wrenching cry. He kept going, telling me all the time that the woeful bellowing was just a nervous reaction. He prised off the bud with the iron and then cauterised the wound with the side of the iron. He had burnt right down to the skull. Little open circles of nothing but bone... Then he did it all again on the second bud.

He gave them each an injection, not a painkiller but something to bring them back round minutes after having their heads burnt. They lay there in a stupor, heads down, looking totally miserable.

At the time I hadn't heard the term 'small humane farm', but I guess this was it ... a small humane farm run by someone who loves animals. I would have argued with anyone who suggested I wasn't caring for these animals. I was doing it right. Some things were hard but I was doing what had to be done. I never questioned what I was doing to these animals ... these babies. I never asked myself if it was morally right. This was the best way. I couldn't see another 'best' way.

10

A Mouse, Six Cows and a Shed Load of Sheep

Wild forebears of the domestic sheep naturally gave birth to one lamb but through selective breeding our domestic sheep usually now give birth to twins ... double the profit. But often the ewe will give birth to triplets and, as she can't generally feed three, she will reject one of them – usually the smallest, weakest one. This is what had happened to Mouse.

Our friendly farmer came round one day with the smallest lamb I had ever seen. He was half the size of an average newborn, just a tiny bag of loose skin. We were told he probably wouldn't survive but if I wanted to try bottle-feeding him I could.

Of course I would try.

The farmer handed Mouse over to me. The lamb was really cold. He had not an ounce of fat on him and no mother to keep him warm. Tucking him inside my jacket, I took him into the kitchen, sat in front of the stove, cuddled him and rubbed him with a towel. It seemed to stimulate him. I thought of Lola crazily licking little Mandy; she had known what she was doing. The more stimulation the better because it gets the lamb going.

I fed him colostrum milked from one of the goats, set him in a box in front of the fire with a hot-water bottle and he fell asleep. After a few more feeds, a few more cuddles and a few more rubs, he was up on his feet with tail wagging, calling for more food.

Little Mouse was getting stronger, feeding well and starting to fill his skin. He soon seemed healthy enough to join the other bottle-feds and he revelled in their company. He was happy and lively.

At a few weeks old, he started going downhill. He wasn't feeding properly and spent most of his time lying down. I brought him back into the house and tried everything I could think of but it seemed that his little system had done enough. His outer body was getting stronger but his internal organs couldn't keep up. He grew weak again, unable to stand. Mouse was a fighter but he could fight no more.

I was holding him as he slipped away. He was a creation of selective breeding. He should never have been born but he was. The wonder is not that he died but that he lived.

∞

We were already tight for land even before all the new babies. Now we were struggling and, come winter, we would need to buy in a lot of animal feed. We decided to look for a bigger place. It actually all happened really quickly; someone was interested in our little smallholding and, at the same time, some other folk were selling their small farm. Within a few weeks, by early summer, we had moved into a farm with lots of outbuildings and sixteen acres.

The previous owner had six female cows – heifers. Their plan, before they decided to move, was to grow the young cows through the summer and then send them to market in October. They asked us if we could keep the girls till then and send four to market; in return, we could have the other two.

Could we do that? Of course we could. We now had cows.

∞

A few weeks after lambing, adult sheep need to be sheared. In their wild state sheep lose their fleeces naturally but now, due to selective breeding, sheep produce bigger fleeces. We have also engineered them to retain their fleeces longer so that we can take them and use them.

The previous year when we only had Lassie, we had sent her back to the farmer for the day and she came back shorn. That was the plan again this year with our three girls but this time we would go with them to watch so that we might do it ourselves in future.

Once shorn, the sheep would be susceptible to all but mild, dry conditions. Too hot, too cold or too wet and they would feel it – and Orkney weather, even in summer, isn't the most dependable. This day was quite cold.

I was designated the role of catcher. The adult sheep had been driven into a barn. They were nervous and upset because they had been separated from their lambs, lambs who were desperately running around the field outside calling for their mums. I would catch a sheep, one handful of fleece at her neck and one handful on her bum, then manoeuvre her through the door to the shearers. I noticed that some of the farmhands were catching them by an ear and a handful of fleece. I didn't join them.

Once handed over, the shearer used handfuls of fleece to flip the sheep onto their bums. They were using electric clippers and seemed to take pride in how fast they could shear, quickly throwing the sheep from side to side like sacks of potatoes. Some of the sheep found it hard to stand once they were released. None of them escaped without a few cuts from the electric clippers and some cuts were much worse than others – much worse.

The farmer's ewes were released back into the field and reunited with their lambs. We loaded our three girls into the Land Rover and took them home. We kept them in the barn that night, together with their babies; it was too cold outside. I treated their cuts with antiseptic spray, gave them plenty of hay and water and said sorry.

11

Monster Chicks and Sheep Dipping

Our idea of hatching some goslings to sell had been put on hold when Onion was killed and Sage had died, so when I noticed one of the hens going broody, I put her and her eggs in an upturned tea-chest with a little run. In the evening, when she was snuggled up and dozing, I slid my hand underneath her and exchanged all her little chicken eggs for two massive goose eggs.

Sometimes I think chickens are really clever and other times I think they are a bit daft. This, thank goodness, was one of this chicken's daftest moments because by morning she was all fluffed up, engulfing the mammoth eggs in her feathery coat. She must have thought, 'Well, I don't remember laying those!' and was probably quite impressed with herself.

She sat on them patiently until they hatched.

We noticed the broken shells first but it was another day before we first saw the two goslings. They were dry, with grey and yellow downy fluff, and were a good size. The hen dropped her wings to shelter them, one on each side, to give them their own little refuge.

I put their box and run onto good fresh grass and put in grain for the hen and a mash for the babies. The chicken, in good mother-hen mode, took individual grains and broke them up with her beak, talking to the chicks all the time. Peck, peck, 'Chook, chook, chook, come on babies, this is how you eat grain.' The goslings, however, naturally preferring to rip up and eat grass, ignored her.

'Chook, chook.' Peck, peck. 'No, like this, chook, chook.'

Rip. Rip.

She started getting agitated because they were paying her no attention.

'CHOOK! CHOOK!' Peck, peck. 'NO! We eat grain. Like this...' Peck, peck.

Rip. Rip

She tried, she really tried, but her babies looked so happy eating the grass, that eventually a very confused mum started eating the grass too.

By the next day they had sorted themselves out. Mum ate the grain: 'Mmm, this is really nice. Are you sure you don't want to try it?' The goslings ate the grass, and they all ate the mash.

They grew really quickly. Each day Mum's wings lifted higher and higher as the goslings sheltered underneath them. Eventually they were bigger than their chicken mum but they still slept one on each side at night, and she would reach up and lay a wing over each of them. They were still her babies.

Once again our farm was full of youngsters ... and there was another baby on the way. I discovered I was pregnant. My baby was due in the spring with the lambs.

∞

At the time, sheep dipping was compulsory to protect the ewes from external parasites. The farmers routinely left the sheep for three or four weeks after shearing to allow time for their wounds to heal. We had no sheep-dipping facilities so once again a friendly farmer came to the rescue; we could dip our sheep along with his. He lived not too far away so we quietly walked our little flock along the road, letting them munch on the verge as they went.

We guided them into a pen at one end of the sheep dip. They knew something was about to happen and started panic bleating while jostling nervously for the position furthest away from the strong-smelling dip. We were told to stay well back at the dry end. The guys who were involved in the dipping were in full waterproofs, some with thick rubber aprons on top, thick rubber gloves, wellies or waders and eye masks; some had masks over their mouths as well, and all had their hoods up. I remember thinking, 'They really don't want to get wet.'

I found out later that the potent chemicals used in sheep dip are classed as hazardous substances and the protective clothing was to stop splashes getting onto the skin or into the eyes or mouth. Without sufficient protection, farm workers could suffer from flu-like symptoms, burning when passing water, dizziness and blurred vision, which could last for weeks ... and that was just from getting some dip splashed onto their skin.

The ewes, however, were pushed down a ramp and into the cold dip. The floor disappeared from under them and they had to swim. Already panicking, they struggled to keep their heads above water. One of the farmhands, using either a broom or his foot, ducked them completely under the water and held them, only for a second or two but it seemed like forever. The sheep came up, spluttering for breath, eyes wide and desperate, struggling to get a footing on the ramp that led them up from the dip and onto dry land where they stood, heads down, mouths wide, tongues out, panting. The substance which the farm workers were so careful not to get on their skin was in the sheep's eyes, ears, noses and mouths. Their fleeces, even though they had only recently been clipped, still held copious amounts of the foul-smelling water and they just stood there ... dripping.

We walked them back, their heads down, our heads down. They didn't stop to eat from the verge. They just wanted home. We had abused their trust ... again.

But these things had to be done... We were smallholders.

12

Cows Go to Market, Pigs, Goats, Geese Go to Heaven

Summer. There were dogs and cats, goats and kids, sheep and lambs, hens, chicks and goslings, cows ... and there was Petunia.

Petunia was now over a year old. She was big; the big pink pig that you probably visualise when you think of pigs. Her run wasn't much bigger than her now, and she was half the height of the dyke. She obviously missed Danish. She didn't play any more; she had no one to play with and no room, but she would put her front feet up on the dyke, look over and call me for a cuddle and a chat. Generally she was very quiet ... except when she came into heat.

Petunia in heat wanted a man and she really fancied my husband. When she saw him she would suddenly jump, lifting her bulk over the dyke like a greyhound, and race down the field after him. She would grunt and nuzzle him, push her shoulder into him and tell him how much she loved him. He had to lead her back with a bucket of feed and then withdraw quickly or she'd be over the dyke again. Every three weeks would see him trying to go about his business, staying low, sneaking around, trying to stay out of sight and down wind.

∞

Summer in Orkney is usually over by the end of August. We had planted and harvested about three acres of hay and three acres of bere barley, a crop grown in Orkney which is well suited to the short growing season. The barley also gave us straw for bedding.

As summer turned into autumn, the bottle-feds were weaned and put in with the rest of the flock. The two goslings, who had now left but not forgotten their hen mother, joined the other two geese, Parsley and Busby.

My little Busby would sometimes climb through the fence to come and see me, leaving the other geese behind. Even when he was pretty much a full-sized gander, he still loved a cuddle. I would sit out of the autumn wind behind a dyke and he would clamber onto my lap, talking his goose talk and telling me all about the day he'd had.

∞

By the end of September the Orkney winds start to pick up. The animals would need shelter. The goats could go in and out of their barn as they pleased, although on bad days I kept them in, but the cows needed to be taken in. We had barn space for them, but our sixteen acres could not grow enough winter feed, not for six, so the four that we had promised to sell for the previous owners had to go.

We hadn't sent anything to market before. I had, of course, seen plenty of cows at the harbour being driven up the narrow ramps onto the boat by farmhands with sticks. At the time I had taken no notice of the racket they were making; cows are just noisy, aren't they? I had never thought that they might be scared, calling for their friends, their mothers. Most of them were the equivalent of children but I didn't know them; they were just cows.

Our guys were only five months old.

We didn't go to the harbour. Our farmer neighbour took them for us. We manoeuvred them into his trailer. They didn't want to go; they had only known their quiet field. They were much bigger and stronger than us, and they could have turned and trampled over us easily if they were inclined to, but they were such gentle beasts that we were able to push them up the ramp without much of a problem.

They started bellowing as soon as they were in the trailer. The ramp was pushed up and closed and they were packed in tight. A head appeared out of the top, eyes wide, confused, scared. The tractor and trailer started forward, which jerked her back down, and they were driven away from everything they knew. They would be herded onto the boat by the men with sticks, together with all the other scared cows. They would stay down in the hold as the boat went around the islands picking up others, then they would be unloaded into small pens at the cattle market. They might stay there for a day, or even two, waiting for the sale.

There would be no friendly faces. They would be driven into the ring and made to move round so that people could see them properly and bid money for them, then they would be driven out of the ring, back into a pen and into another trailer. Now it would depend: these were girls and they might be lucky because the farmer might have bought them as breeding stock. If not, they would go straight to the slaughterhouse.

I convinced myself that they would be breeding stock. In the real world, as breeding stock they would live their lives having baby after baby until they were spent; then they would be sent to the slaughterhouse. But in my little world, in my head, they were going to be fine.

I hoped they would be fine.

∞

The two cows we kept, Ada and Gertie, went into stalls in our barn. I tied them with a thick chain around their necks. They could stand up and lie down, that was all. This was normal practice. Unlike most of the cows on the bigger farms who would stay chained up on concrete floors for the whole of the winter, I let our girls out into the fields on all but cold or stormy days. They would come back in the evening when I called them, knowing there was a big feed waiting for them and a thick bed of straw.

It was that time of year again, time to get rid of the surplus stock before we started wasting feed on them. I had managed to sell the two white female kids as milking goats. Mac, Lucy's male lamb, had gone the same way as the cows. I had given him to our farmer friend as a thank you and he had put him in with the lambs that he was sending to market. Little Mac, whose fate was predetermined before he was born, would have gone from the cattle market straight to the slaughterhouse. People like to eat lamb.

We were killing – small kills – all year round now, taking a young rooster's life every time we fancied a bit of chicken. Now for the big kill. We only had Petunia, the pig, and Jezz's male kid. That didn't seem very much and we had a large freezer... so we killed Jezz as well.

I wanted to breed British Alpines so Jezz, a Saanen, and Schnooks, a Golden Guernsey cross, were no longer in my plans. Jezz was five, older than Schnooks, and not giving as much milk. It made sense. I led her round to the barn. She trotted happily by my side, wondering where we were going. I remember feeling bad about it, but what else could I do?

No one would buy an 'old' goat with one horn.

So we had a large pig, a large goat and a kid all chopped up and bagged in the freezer. We had chicken whenever we wanted it, eggs, milk, butter and cheese. I had even made potted heed. Not only that, my husband would bring back wild duck, rabbit and pigeon. We had protein covered. We would probably survive.

Christmas was coming and people had ordered geese from us for their Christmas dinners: Parsley and the two goslings. A week before Christmas, my husband took an axe to their necks. I plucked them, pulled out their innards and handed them over, clean and ready for the oven. That was how the customer liked to receive them. Now it was Busby's turn; he was a fine gander, but he had always been destined to be Christmas dinner. I called him and he came running over. My husband picked him up.

If you tuck a goose under your arm, they will reach out their heads straight forward, lengthening their necks. Busby lengthened his neck out ... over the chopping block. I turned my back, closed my eyes and covered my ears.

Happy Christmas.

13

Different Childbirth Experiences

By spring, Alex's kid Jamie had grown into an impressive male, my British Alpine stud male. He was big; if he reared onto his back legs, he was over six feet tall and, as he loved playing, he reared up often. He would dance around up there for a few steps and then come whamming back down, head in butting position, a little way off at first and then, once he'd decided that I was up for a game, he would aim at my kneecaps, stopping about two inches away. I didn't have to move; he never hit me and I trusted him completely. He would do it over and over again until finally settling down for a neck scratch and a cuddle.

Our land led down to some dunes and, on the other side of them, a beautiful white sandy beach. Jamie loved coming to the beach with me and the dogs. As I collected seaweed for the goats, he was just another dog running in and out of the water, playing chasing games and sometimes giving me a wee push so I wouldn't forget he was there. He loved his trips to the beach and we spent hours there by ourselves, just me and the dogs and Jamie.

Obviously Jamie couldn't mate with his own offspring, or his mum, so in the autumn I had bought another young male and another female. They were both British Alpine and both around six months old. On top of that, I had been given a part bred BA female in milk, Cressy. The little BA herd was coming on nicely.

Schnooks, Alex and Cressy were all in kid and all of the ewes were in lamb. It was going to be a busy spring.

I was the first to produce. Wow! Suddenly I appreciated what those ladies went through every year ... giving birth is really hard!

We had a beautiful baby boy. I loved him from the moment I discovered I was pregnant and when he was born I got to keep him and cuddle him. I knew nothing bad was going to happen to him. He woke every night and early morning, my little alarm clock for checking the sheep.

The ewes all lambed beautifully and made a lot less fuss about it than I did. Some had singles, some had twins. There were boys and there were girls. Some needed a bit of help but I was more confident now and gave help where it was needed. We didn't lose any and I still marvelled at every birth.

The Elastrator came out again and I castrated the males and did all of their little tails. This time I didn't think too much about what I was doing to them, and I didn't watch them in the field after I had done it. I just went back into the house and cuddled my son.

Another smallholder, who knew I had an Elastrator, asked me if I would do their lambs. Word quickly spread and that spring I found myself mutilating quite a few lambs with those tight little bands.

Alex, who had been put to a British Alpine male, had two girls. Perfect: two more to add to my herd. Schnooks and Cressy had been put to an old billy on the island; it didn't matter about their offspring because they wouldn't be pedigree. We didn't want the kids, we just wanted the milk.

Assuming Schnooks and Cressy each had twins, four goats in the freezer would be too many; our freezer was already full to the gunnels. So, when Schnooks started showing signs of kidding, I called my husband. It would be best to kill them straight away; I wouldn't get attached to them and we wouldn't be wasting money on feeding them. I didn't even want to see them. That would be too hard. Out of sight, out of mind.

I hid in the kitchen for two hours before my husband came back in. It was done. I didn't ask, but I knew how he had done it and this was confirmed when I went into the barn and saw the bloodstained hammer lying on the floor.

Schnooks was calling for her kids. She had lost little Tyler after her first kidding; the previous year her kids had been taken from her when they were just a day old ... and now this. I tried to comfort her. She hadn't even got to see them this time but she would have smelt them, their smell would have been everywhere. I buried my face in her neck and we cried together. Was her milk really worth this?

Cressy kidded later that day, a boy and a girl. Although they were far from pedigree, they both had the British Alpine markings.

There was no use for the boy so he would have to go into the freezer – but maybe I could keep the girl ... or sell her?

The next day I took the babies away from Alex and Cressy. Now there were three grieving mums and four distressed kids. The noise, the frantic calling, was hard to ignore even in the kitchen. I couldn't imagine what it would be like to have my baby taken away but... Harden up, Vicky. You want the milk.

A week later, the three female kids took a trip to the vet and had the top of their heads burned down to the bone. I had become used to some things but I never got used to that.

14

Just Things You Have to Do

I was bottle feeding four kids and one lamb, and looking after my son. The government gave me one tin of baby milk powder a week which, as I was breast feeding, I didn't need. I made up the milk and gave the goat kids the human-milk substitute while we drank the goat's breast milk... Strikes me as ridiculous now.

After our lambing had finished, our friendly farmer came along with another lamb for me to bottle feed. This one was the smallest of triplets; his front feet were tucked under at the first joint, so he looked as though he was walking on his knuckles. His mum had rejected him but he would not have been able to reach the udder anyway. I took him.

There was a gorilla in the news at the time whose name was Gay. The name suited this little chap too, not only because of his knuckle walking but because he was always so happy. His head was black, with ears that kind of stuck out to the side rather than up. His little legs were black and his coat – dark grey, short and quite curly – was two sizes too big. He was loving and playful, even though he was stumbling about on his knuckles.

I kept him in the house. The other bottle-feds would have been too boisterous for him. He was small and I was determined not to lose him like I had lost Mouse, so Gay stayed in the house for longer. I gave him more cuddles and massages. I straightened out his little feet and encouraged him to walk on them. I was Mum to all of the other bottle-feds but they had each other and played happily when I wasn't about. Gay, however, relied on me completely. He didn't even know he was a sheep.

As he grew stronger and his little feet straightened, I put him out with the others but he never mixed. He stayed at the closest point to the house or to me. Out of the field, he followed me around everywhere. His voice was loud for such a wee chap and he used it

whenever I disappeared around a corner...
'Maa, maa.' It really sounded like he was
calling for his ma which, of course, he was. I
wanted him to be a sheep, but he just wanted
his mum ... me.

When he was small he loved being picked
up and cuddled, but as he grew bigger I had to
sit on the ground so that he could clamber over
me and snuggle his head over my shoulder.
There he'd stay, chewing the cud and burping
happily into my ear.

∞

Summer came and so did the shearing. We
decided to do it ourselves with just the help
of another smallholder. It would be fine. We
would be nice. We had seven ewes to shear.
I would be catcher again and the men would
take their time with each one to avoid nicking
them as much as possible.

The problem with taking more time was
that the ewes became even more stressed.
They could hear their babies calling for them
from outside. No matter how nice the men
wanted to be to the girls, they still had to
restrain them and force them into unnatural,
uncomfortable, vulnerable positions and hold

them there. The longer they held them, the more stressed they became.

At the end of each shearing, when the guys let them go, the normal reaction would be for the sheep to run away as fast as possible but some of them just stood there like they were in shock, heads down, panting. I seriously feared that one of them would have a heart attack.

It took us most of that day to shear seven ewes. The farmers would never have taken that much time or that much care. Even if they wanted to, they just wouldn't have had the time.

∞

In Nature, animals get their B12 from drinking out of rivers and streams and from eating really short grass. Basically B12 is from dirt, but as most farm animals are fed hay and grain when grass is in short supply and drink tap water from troughs, they don't naturally get the dirt to make B12 so they need a supplement. To give them this, I had to restrain the animal (goat, sheep, cow), slide the long metal 'gun' over their tongue and down their throat, and squeeze the trigger which released a huge tablet down the throat. I had become quite proficient and a number of smallholders asked me to do their animals as well. The vet wormed them all using a similar method.

To be truthful, the vet was a bit nervous of bigger animals, especially horses. Once he found out that I had a background working with horses, he asked me to help by twitching any horse that he had to work on. Twitching is the normal practice for restraining a horse. The twitch is a stick with a rope loop attached to one end, we used them when I worked at the stables. I used to pull about half of the horse's upper lip through the loop and hold it there as I turned the stick, twisting the loop and tightening it around the lip. The horse would throw its head around but settled very quickly because movement, when being tightly held by the upper lip, was too painful. I cringed the first time I saw it done but I was assured (by people, who had never had it done to themselves) that it didn't hurt; the lip would just go numb. I guess it was similar to a tight band around the testicles ... eventually it goes numb.

So I became the go-to person for castrating, giving B12 supplements and twitching horses. Anything they didn't want to do, I would do ... apart from killing. I could never do that.

15

From Smallholder to Small Farmer

After the shearing came the dipping. It was clear that the ones who had been dipped the previous year remembered the experience and they seemed to pass the message onto the others because they were all really nervous.

This time it wasn't the easy walk down the road that it had been before. As we got closer to the farm, they started trying to turn back. They must have smelt the dip. A couple broke through the roadside fence into a field and ran but, as much as they wanted to go back home, they also did not want to be separated from their flock. They stood a distance away, calling to the others. This, of course, upset the rest of them and they all started looking for a way through the fence.

I jumped into the field and circled round behind the escapees but they were not for going back. The farmer, who had seen we were having trouble, sent over a couple of farm hands and a sheepdog and, with lots of running and arm waving, we managed eventually to get them up the road and into the pen.

Then there was no escape for them. They were forced into the dip and submerged as before.

∞

Having cows had never been part of our plans but now we had them, we loved them. The two remaining girls, Ada and Gertie, had proved to be gentle, loving animals that were easy to look after and loved company and cuddles. The lambs that we had sent to market had only brought us a few pounds but a young cow would sell for much more. If we sold three young cows each year, that would help greatly towards the cost of goat mix and powdered milk. So, we decided to get another cow.

An adult was beyond our budget but we could pick up a day-old calf pretty cheaply from a dairy farm, where all of the male calves and most of the female calves are surplus to requirements. We ordered a day-old female from mainland Orkney. She would have been taken from her mother, crated up and put straight on the boat, lonely, confused, very scared and wanting her mum, but we didn't think about that. When she was unloaded we just saw the cutest little black-and-white calf, still wobbly on her legs, who wanted to suck my fingers.

Her name was Lily. Once home, I made up some powdered milk in a bucket. She had no idea what it was; she was looking for a teat to suck or something similar. I put my fingers in her mouth and she sucked desperately, her little tongue wrapping around them. I lowered my hand into the milk so that she got some milk as she sucked.

This became routine. Initially I left my fingers in there as she drank but soon I could get her going with my fingers and then take them away to leave her drinking from the bucket. After a couple of days, she was drinking straight from the bucket. I fed her little and often so that her small tummy didn't get a fast influx of milk which might cause problems. To me, she seemed quite happy drinking her milk from a cold plastic bucket instead of from her mother's warm udder.

I was soon able to put her out with Gertie and Ada and they mothered her immediately, taking it in turn to nuzzle and lick her. She grew fast and strong and became as playful as my little bottle-feds ... just a lot bigger.

∞

Spring turned to summer. We bought another piglet, Marigold. We planted more bere barley, more tatties and grew fields for hay. We had sixteen acres, plus the use of some common land. A fairly small vegetable garden gave us more than enough vegetables for the year and the rest of the land went to grazing animals or growing feed for them.

We still had to buy in some feed, as well as supplements and wormers, and we had vets' bills to pay. We started selling vegetables; we had plenty and they sold well. Come autumn, we would have four lambs to sell ... and Gay.

We actually sent the four lambs to market towards the end of summer. A bit early, but they were well grown and the consumer likes young lamb so they would sell easily. Again we grouped them in with some of our neighbour's lambs.

The evening before they were to be shipped, we separated them from their mothers and shut them into our neighbour's barn. It was an early boat and I was still in the kitchen when I heard his Land Rover and trailer go rumbling past. 'Bye, babies.' I hoped they'd be fine, but I didn't give it much more thought. Out of sight out of mind.

The cheque came in the post a few days later.

16

Gay

Gay hadn't gone with them. He was still a bit small. He spent most of his time not grazing but following me, calling for me or looking for me. He had never mixed with the others, spending more time with the dogs than with the sheep. But when autumn came he had to go. He was a castrated male; there was no use for him on the croft. I knew when I took him on that this time would come. I couldn't keep him through the winter; consumers wanted baby lamb, not year-old lamb.

So all the other lambs had gone and now there was only Gay. I lifted him into the back of the truck and got in with him. He snuggled in, tucking his small black nose between my knees and spreading his little feet to steady himself. He had never been in a Land Rover before and the road was rough, the suspension non-existent.

We arrived at the harbour and were told to put him into a small paddock where all the other lambs were waiting to be shipped. I didn't have to drive him in, he just followed me. I squeezed out of the gate and shut it quickly, closing him in. He started immediately, 'Maaaammm, maaaammm,' running up and down the gate. I turned and walked away. Don't look back.

We went over to the harbour office to fill in the form.

'He won't herd with the others,' I said.

The harbourman looked at me. 'He's a sheep. He'll be fine.'

'He doesn't know he's a sheep.'

So, we could go now. The harbourmen would board the sheep but, although I really wanted to get out of there, I couldn't. There was a lot of noise: the boat's engines, vehicle engines, cows and sheep all calling for their mums. Even with all that background noise, all I could hear was Gay calling for me, 'Maaaammmm.'

I couldn't leave. I needed to know he was okay. As we left the office, I saw the lambs being driven down the pier towards the boat. 'Oh!' I cried out. There was Gay at the back, a rope round his neck, being led like a dog by the harbourmaster.

They had opened the gate, all hands ready to drive them, and Gay had made a break for it. He had dodged between their legs and run off down the road looking for me. It had taken a number of them to catch him and now they pushed him into a pen with the rest of the lambs and started driving them all up a ramp and into the bowels of the boat.

By this time I was distraught. I couldn't look and hid behind a wall.

'Maaaammm, maaaammm.'

Hands over my ears, I peeked around the corner, sobbing uncontrollably. Gay was facing the wrong way, trying his hardest to go against the flow but getting pushed backwards up the ramp. He was still calling hysterically, 'Maaaammm, maaaaammm.' I watched him as the flow of sheep carried him into the boat. The ramp was pulled away and the door shut.

All of the lambs were scared. All of them were bleating, crying out, but from inside the boat above the sound of the engines, I could still hear Gay. I could still pick out his voice. 'Maaaaammm.'

17

Some Get to Live, Some Get to Die and I Get to Thinking

There were new incomers on the island and I managed to get in quick and sell them Cressy's female kid. But more incomers meant more goats, which meant more kids, and everyone was now trying to sell them. Everyone wanted the milk but there wasn't much call for the kids.

I now had five pedigree British Alpines, one crossbred, and Schnooks. It was October, decision time ... put Schnooks in kid, kids that would be no use to me, or put her in the freezer?

I had seven ewes, plus four female lambs coming on. Did I put all seven to the ram or had Lassie served her time? Was it time to put her in the freezer?

Schnooks, the Golden Guernsey, Lassie, the 'old' ewe and Cressy's male kid all took the long walk to the barn. Their 'thank you' was a bolt in their heads and their throats slit. They were skinned, chopped up, bagged and chucked into a freezer that was already full to bursting. This killing made me sick to my stomach. There was 'being hard' and 'being hard' and I wasn't quite making either of them.

_∞

One night I had a dream, a dream so vivid that to this day I remember every detail. I had gone into a cave. There was a dog with me. He wasn't my dog; he wasn't a dog that I knew, he was just 'a dog'. There was a huge rock fall outside which sealed the entrance. We were trapped. No one knew we were there.

I sat on the ground with the dog. I could kill and eat him, which might keep me alive until someone found me, but I didn't. I knew in my dream that the dog would survive, so I sat there knowing that I was going to die and I was happy, content to know that the dog would live.

I woke up still happy. Thereafter the dream would often come to mind, so vividly as if it really meant something. I started to question my new morals. 'New' because these weren't my morals when I was that little girl with all her pets who didn't want to eat animals. My upbringing, my whole life up to that point, had taught me that you had to kill animals (or have someone else do it for you) to eat ... and you had to eat.

I'd heard of vegetarians; most people thought they were weird, cranks. And I had actually met a couple of vegans before I knew what a vegan was. They were strange, hippy people. I didn't talk to them.

But were my ideas, my beliefs, changing now? Could there be another way?

∞

Marigold, the pig, had escaped that slaughter session. A couple of weeks before the killings, she became poorly. She developed a rash over most of her body and was obviously not well. By the time the vet came, she was lying out full length on the ground. The vet and I squeezed through the low door that led into her small barn. On the other side was an opening leading out to her run, which was just big enough for her to get though. I could have crawled through it but the chunky vet had no chance.

Marigold was quite big by now. She didn't move as we entered; she didn't even welcome us with a grunt. The vet who, you will remember, was a bit scared of animals, said that she had measles and she needed antibiotics.

He produced a needle from his bag and I asked if I should hold her while he injected her. 'No,' he said. 'She's far too poorly to notice.' He stuck the needle in her back leg and, in an instant, she leapt up squealing and spun round.

The vet, who was not an athletic type, also leapt up and was out of the door in a flash, pulling it closed behind him. If Marigold had suddenly become a killer pig, he had just shut me in with her. She turned back to me, grunting her low, complaining grunt. I rubbed her head and chin and told her it was alright and she settled back down for a tummy scratch.

I was smiling to myself as I left the barn and found the vet waiting nervously outside. He gave me antibiotics for her and told me that she would not be fit for eating until at least six weeks after the rash had gone. So, Marigold got to live ... for a while longer.

18

Going Vegetarian

Another Christmas came. No goose that year but we had pork (Petunia) and leg of goat (Schnooks) and for the next week we stuffed ourselves. It's called festive feasting. By New Year we were bloated and lethargic.

Most new years see a flush of new diet and healthy eating books. This year there was a book by Leslie and Suzanna Kenton called *Raw Energy*. The diet, a ten-day fast, involved eating nothing but raw vegetables and fruit. A bit extreme, but we were in need of something extreme and we had plenty of veg, so we tried it. I lost ten pounds in ten days, felt absolutely fantastic and full of energy, and I felt 'clean' – better than I had felt in a long time. However, after the ten days habitual behaviour kicked in and we thought we were ready for a feast again.

I asked my husband what he would like and he replied, 'A big leg of mutton.' So I roasted one and presented it on the table. We both looked at the leg, glistening with fat. We didn't see a leg of mutton, we saw Lassie's severed leg. I felt sick. Were we going to eat that? Suddenly we weren't hungry.

We literally went vegetarian then and there.

There was the small matter of a freezer full of meat. We gave some away and used the rest as dog food – Schnooks, Lassie, Jezz, Petunia. The local farmers thought we had gone crazy and worried that we were going to bring up our poor son vegetarian. How would he survive? I caught one sneaking a spoonful of beef mince into my son's mouth. He took it, tasted it and then spat the whole lot out again...

We were now vegetarian farmers. We were not going to eat any more of our animals, but we still had a farm.

Ewes, goats, cows; they were all pregnant. Female offspring we would probably keep but what would happen to the males now?

∞

Six weeks had passed since Marigold's rash had cleared up. She was back to full health. She was eating a lot and we were no longer going to eat her. She needed to go; she was a good size, ready for slaughter. We sold her to a farmer just a few fields away.

Two days later I was outside. It was a bright, still day, one of those days when you just enjoy the quietness. Coming from across the fields I started to hear squealing, broken at first, but quickly becoming long and loud until it filled the air. It sounded like a human screaming in agony. My heart raced. It seemed as if it went on forever.

Marigold.

'I'm so sorry.' I covered my ears and hurried back inside.

∞

Spring had arrived and the lambing started, beautiful lambs being born. Being vegetarian didn't change the fact that I needed to castrate and dock them. Gertie and Ada produced two beautiful calves, Winnie and Horace; on top of that, we had ordered another day-old calf from the mainland, this time a little boy, Bertie. We would grow him on and sell him.

When he arrived on the boat, Bertie was smaller and even more adorable than Lily had been. He was the cutest calf you've ever seen, a chocolate-brown teddy bear.

Within a few weeks Horace and Bertie needed to be castrated and ear tagged. I could do the ear tagging, a special tool crunched a hole in their ears and left a big plastic number tag, but I wasn't prepared to do the castrations so I called for the vet.

The two boys were in the barn, tied up short. The vet injected them around their delicate area (I had insisted on local anaesthetic) then, taking out a scalpel, he slit the scrotum, pulled the testis out individually and, with a sharp jerk, ripped them out and threw them on the floor. The calves bellowed, stamped and fought to get away but they were tied firm. They had obviously still felt something. Had the anaesthetic even worked? The vet gave them a quick spray with antiseptic and they were done. He had obviously done it many, many times. I was left with instructions to keep an eye on them in case they got infected, but I wasn't sure if either of them would let a human near them ever again.

There were a couple of ewes left to lamb and the three goats to kid. My husband was away when Cressy decided it was her time. We wanted the milk but we didn't want the kids, not even to eat now, so we had decided that her kids should be killed at birth just as we had done with Schnooks the previous year. Only my husband had done the killing then; now there was only me.

I knew what to do. It wasn't physically hard. As soon as they were born... It was best. It's what dairy farms do all the time with the babies that they don't want. I needed to be tough. Switch off emotions.

I cannot bring myself to tell you what I did. It was awful ... more than awful. I have no words...

What sort of person can do that? It's so easy to let someone else do it for you but when you see it ... when you do it...

I did it, all the time weeping uncontrollable sobs. I could hardly see. As soon as it was done I wanted to undo it. Go back in time. Make it never happen.

I cried for days.

I still cry now.

19

Goodbye Orkney

I wanted to leave this place. This was not why we had moved to Orkney. If we stayed there would be more killings; being vegetarian wouldn't stop that. I realised now that if we stayed I would be putting little Bertie on the boat, sending him up that ramp just like Gay. There would be more castrations, more disbuddings, shearings, dippings, more animals sent to market, sent to slaughter. I had never wanted to hurt an animal, yet here I was killing them. What was I doing?

Where was the picture book of my childhood? Happy animals living out their lives in green pastures with the happy farmer who would never cause them harm?

We sold the farm. I sold all the animals

to other farmers and smallholders, pleasant, friendly people who believed that they were doing the right thing just as I had. All of the animals, my animals, would have ended up being slaughtered sooner or later, either at the farm or in a slaughterhouse. All of them would have ended up hanging by their back legs having their throats slit. Bertie and Lily, Ada, Gertie, Lola, all the sheep, even pedigree goats, still end up being slaughtered. If they were lucky they would have been stunned first. None of them would have lived long and died of old age.

They were farm animals.

I didn't want to be around to see it.

∞

We left Orkney and moved to a normal life in Inverness; a normal life, where you pick up your milk from a shelf in a supermarket.

Thirty Years Later

'After this there is no turning back (you take the blue pill) the story ends, you wake up
in your bed and believe whatever you want to believe.
You take the red pill – you stay in Wonderland and I show you how deep the rabbit hole goes. All I'm offering you is the truth.'
The Matrix

I was no longer eating meat and I no longer had to consider where the milk or cheese or yoghurt came from, or the process of getting it. I could just pick it up. I could buy my eggs packaged up in a nice little box with a picture of a happy chicken on the lid. Everything was normal, convenient, easy. I didn't have to kill anything. No one was getting hurt. I believed the marketing; I believed what 'they' wanted me to believe. My life now was a million miles from my Orkney life.

And then one day, without even trying, I took the red pill.

The red pill for me was a picture of a slaughtered dairy cow hanging by one back leg having the still-living foetus of her baby cut from her womb to hit the cold concrete floor and die there. It was a world that I had chosen to forget.

I had a Facebook 'friend' who regularly put up graphic pictures of milk cows in factory farms and slaughterhouses and pictures of day-old chicks being thrown into macerators so that we can eat eggs. I wouldn't believe it. This wasn't happening in this country, in this time. Or they were just cruel, isolated events. After all, our farms and slaughterhouses are Red Tractor/RSPCA Approved.

I deleted her, unfriended her. I didn't want to see those pictures, they spoiled my day.

I don't know how that one picture slipped through onto my page but there it was. There she was, a cow, a mother, dangling from a chain, in this country and in this age, and she wasn't alone. Every year of her life she had given birth, had her babies taken away and given gallons of her milk for us; now she was spent, only good for cheap burgers an dog food.

It hit the spot. I swallowed the pill. I saw. I knew. I finally got it. No animal should die for me. Not for meat. Not for dairy. Not for eggs. No animal should be in pain or misery ... for me.

It had taken me sixty years.

∞

Now I'm vegan, I finally feel like I'm being true to who I am, true to the person I was born to be. I'm living my values. I can, without fear of judgement, love and respect every living thing that has evolved on this planet, living, sentient beings that are here with me, not for me.

Farming had felt like a natural, wholesome thing, but it was a world where if an animal was of no 'use', if he or she didn't earn their keep, you just killed them. A world where you got the most out of the animal when they were alive and then ended their life because now you were wasting money on them. And I had thought it was okay. It was normal, it was how it had always been done, so it must be alright. How stupid was I? How blinkered?

I killed two kids, two newborn babies, and

it wasn't against the law. In fact it was – it still is – normal. I killed the kids so that I could take their mother's milk, milk that I didn't need, milk that none of us need. Later, when I moved away from Orkney, through my purchases I still paid for babies to be taken from their mothers and killed ... and it was normal. It still is.

I will never forgive myself for what I did to those animals on that little farm. I will never understand how I didn't see what I was doing back then, how I didn't 'get it'. But I will make sure that I never willingly harm an animal again.

So I wrote this book:

- for all the vegans who can't understand why they didn't get it sooner
- for the meat eaters and vegetarians who still don't get it
- for the small, free-range, organic, humane farmers who think they've got it
- and for the animals. I'm so sorry I didn't get it, but I've got it now.

One day the whole world will get it.

Useful informtion

Going vegan: Try it with brilliant, free support
challenge22.com
30dayvegan.viva.org.uk
veganuary.com

Must sees:

The best speech you will ever hear (You tube)

What the Health (Netflix)

Cowspiracy (Netflix)

Earthlings (Earthlings.com)

Forks over Knives (Netflix)

The Game Changers (Netflix)

Acknowledgements

It wasn't easy writing this book, bringing back the memories, and without the support and encouragement of my amazing family: husband, Robin, and children, Mike, Lana and Finlay, this book would not have been written.

From reading first drafts to answering endless questions like, "Does this sound alright.....?" or, "How do you spell.....?" they supported me all the way. Thank you. I love you all dearly. I promise I'll start cooking meals again now.

And I have to thank everyone at my publishers, 2QT Limited: Catherine, for guiding me through every step and making the whole publishing process so easy.
Karen, for her awesome editing skills and for never getting fed up with, "Just one more change....." And Charlotte, for the lovely cover design. Thank you all.